Evelyne Mann

Validation of deep sequencing data from Drosophila mauritiana

AF167897

Evelyne Mann

Validation of deep sequencing data from Drosophila mauritiana

by classic Sanger sequencing

Reihe Realwissenschaften

Impressum / Imprint
Bibliografische Information der Deutschen Nationalbibliothek: Die Deutsche Nationalbibliothek verzeichnet diese Publikation in der Deutschen Nationalbibliografie; detaillierte bibliografische Daten sind im Internet über http://dnb.d-nb.de abrufbar.
Alle in diesem Buch genannten Marken und Produktnamen unterliegen warenzeichen-, marken- oder patentrechtlichem Schutz bzw. sind Warenzeichen oder eingetragene Warenzeichen der jeweiligen Inhaber. Die Wiedergabe von Marken, Produktnamen, Gebrauchsnamen, Handelsnamen, Warenbezeichnungen u.s.w. in diesem Werk berechtigt auch ohne besondere Kennzeichnung nicht zu der Annahme, dass solche Namen im Sinne der Warenzeichen- und Markenschutzgesetzgebung als frei zu betrachten wären und daher von jedermann benutzt werden dürften.

Bibliographic information published by the Deutsche Nationalbibliothek: The Deutsche Nationalbibliothek lists this publication in the Deutsche Nationalbibliografie; detailed bibliographic data are available in the Internet at http://dnb.d-nb.de.
Any brand names and product names mentioned in this book are subject to trademark, brand or patent protection and are trademarks or registered trademarks of their respective holders. The use of brand names, product names, common names, trade names, product descriptions etc. even without a particular marking in this works is in no way to be construed to mean that such names may be regarded as unrestricted in respect of trademark and brand protection legislation and could thus be used by anyone.

Coverbild / Cover image: www.ingimage.com

Verlag / Publisher:
AV Akademikerverlag
ist ein Imprint der / is a trademark of
OmniScriptum GmbH & Co. KG
Heinrich-Böcking-Str. 6-8, 66121 Saarbrücken, Deutschland / Germany
Email: info@akademikerverlag.de

Herstellung: siehe letzte Seite /
Printed at: see last page
ISBN: 978-3-639-49133-3

Acknowledgement

Many thanks go to Univ. Prof. Dr. rer. nat. Christian Schlötterer for providing me the opportunity to do research in his outstanding laboratories. I appreciate his patience and cooperativeness. He introduced me to the fascinating world of research. I gained varied and wide experiences during the work on this thesis.

Special thanks go to Dr. Daniela Nunes for supervising me. Her way of support in methodological questions and in discussing troubles have always motivated me and helped me to benefit as much as possible from the work.

I want to give my thanks to all members of CS-lab, especially all PhD students, for their competence and dedication to help me in research. Thank you for all advises as well as for the constructive discussions.

I owe thanks to my family who unremittingly supported me during my years of study. They made this work possible. Special thanks go to my dad, who always supported me in times of critical need and Andreas, who always kept a sense of humor when I was tempted to loose mine.

Table of contents

Abstract

Massively parallel sequencing technology (MPST) has a wide range of applications. Its use in SNP (single nucleotide polymorphisms) detection is already widespread and promises results of high accuracy. The aim is to validate data generated with the Genome Analyser II (GA II). SNPs, detected by using different mapping and SNP estimation parameters implemented in the bioinformatics tool CLC, were compared to SNPs that are detected by using the method of conventional Sanger sequencing. For my study I used lines of *D. mauritiana*. As reference genome in CLC I am bound to use the genome of *D. simulans*, because there is no available genome of *D. mauritiana* until now. The critical part is to map the raw data of GA II (reads) against a reference genome, because the itemized genome of *D. simulans* is not in high quality and the reads used in this study are short (~50pb). Therefore, reliable mapping is not expected.

Using a stringent parameter set (short parameter set) that allows few mismatches of reads when mapping against the reference genome, only few regions are recovered which show divergence to the reference. 60%-81% SNPs detected with this parameter set are false positive ones compared to Sanger sequencing. If a less stringent parameter set (long parameter set allowing more mismatches) is used this results in a very high number of false positive SNPs. Two times more SNPs were recovered with this parameter set, thereof 70-80% are false positive ones. Based on my results I drew conclude that the best approach for SNP detection is to make one first run permitting a high number of mismatches. This turned out to be necessary to get a more appropriate consensus, because a reference genome of the same species sequenced was not available. The next step should be the use of more stringent values to reduce the high number of false positive SNPs.

1. Introduction

In the last couple of years the increasing number of massively parallel sequencing technologies promised to revolutionize genetic research (Shendure et al. 2004; Gupta 2008; Morozova et al. 2008; Shendure et al. 2008; Guryev et al. 2009; Harismendy et al. 2009), medicine (Ramaswamy et al. 2003; Campbell et al. 2008; Guffanti et al. 2009), biotechnology (Schuster 2008), and many other fields of scientific research. The aims of my study comprise validation of pooled population data generated by Genome Analyser II (GA II), the comparison of different mapping parameters, SNP detection runs implemented in the bioinformatics tool CLC, SNP frequency estimation considering mapping errors and divergence of the reference genome in CLC as well as sequencing errors.

1.1. Massively Parallel Sequencing Technologies (MPST)

The so-called "massively parallel DNA sequencing technology" provides an unprecedented speed in sequencing and advances ahead research with its optimized capacity (Shendure et al. 2008). It took years to detect the whole *Drosophila* genome with conventional sequencing methods. Now massively parallel sequencing gives us the opportunity to gain the raw data of a genome in a few days because millions of sequencing data (reads) can be generated in parallel. Fields of application of MPST in genetics are scanning for undefined mutations (Ahmadian et al. 2006), doing SNP genotyping (Odeberg et al. 2002), using it for transcriptome sequencing to achieve quantification of alternative splicing and gene expression (Wilhelm et al. 2008) and to obtain paired end sequencing to discover inherited variation (Campbell et al. 2008). MPST is also used successfully as a tool in preventative medicine and in cancer research (Ramaswamy et al. 2003). Human resequencing studies take advantage of the newest technologies to understand polymorphism and mutation in human individuals (Wheeler et al. 2008).

1.1.1. The importance of MPST in human medicine and research

MPST offers many new chances in human medicine and opens up new perspectives in clinical diagnostics. In the following I want to outline some successful projects that have benefited from the new technologies. Cancer is defined as a genetic disease caused by an accumulation of mutations, which lead to proliferation and cell growth in uncontrolled dimensions (Vogelstein et al. 2004). Campell and Stephens describe new possibilities in cancer classification based on gene expression monitoring (Campbell et al. 2008). For characterizing diseases (in this case acute lymphatic leukaemia and lung cancer) they needed high throughput and sensitivity sequencing methods like MPST. Chen et al. identified a testis cancer gene (CT45) with massively parallel signature sequencing by doing analysis of mRNA and protein expression (Chen et al. 2009). In the sector of 'human immunodeficiency virus disease management' MPST has proved itself in practice: By sequencing a patient's viral population the mutational status offers remarkable information about the most effective drug cocktail (Kozal 2009). By the end of 2009 researchers want to be able to resequence a whole genome of a mammalian based on a budget of $5000. The aim is to reduce costs as much as possible (<$1000) (Guryev et al. 2009). By applying the method of conventional Sanger sequencing the whole human-genome sequencing of 3 billion base pairs would result in about $300 million (von Bubnoff 2008). If costs are cut in such dimensions, MPST might become a common diagnostic procedure in studying genomic variation. This method could be used to analyse the coherence of human diseases and to show how multiple variants of gene mutations interact with each other (Guryev et al. 2009).

6

1.2. MPST products, Sanger sequencing, bioinformatics tools

The technology of all MPST products is based on simultaneously sequencing of molecules after DNA has been separated. Products which work with new sequencing technologies are GA II from Ilumina (USA), 454 Genome Sequencer FLX Instrument from Roche Applied Science (Basel), SOLiD platform from Applied Biosystems (USA), Polonator (USA) and HeliScope Single Molecule Sequencer technology from Helicos (USA). The GA II uses the method of bridge amplification and base-by-base sequencing. Every fragment is cloned up to ~1000 times, resulting in millions of clusters that can be sequenced coevally (http://www.illumina.com; Shendure et al. 2008). Great advantages of GA II are the extremely high throughput of two Gigabases per run, the low price and good processing of homopolymer repeats (some equal, abreast bases, such as TTT or CCC). A disadvantage is that read length is limited by signal decay or by fluorescent labels which are incompletely cleaved, so read length doesn't go beyond 75 bp (with paired-end reads 2 x 75bp) until now.

The method applied to use the Genome Sequencer FLX Instrument (454 platform) is pyrosequencing. Its benefit is the longer average read length up to ~250 bp with more than 99.5% single-read accuracy (http://www.454.com/products-solutions/system-features.asp). This Genome Sequencer also offers a higher read density in case of accumulation of G and C bases. One of the drawbacks of this method is that homopolymers can't be sequenced accurately from time to time. This limitation is caused by the absence of a terminating residue which prevents multiple incorporations (Shendure et al. 2008). Additionally, pyrosequencing is much more expensive than base-by-base sequencing with respect to per-base-cost.

7

SOLiD from Applied Biosystems uses emulsion PCR to amplify sequencing features. It has a great disadvantage: its short read length of only 35bp. Compared to the GA II costs per base are nearly equal (Shendure et al. 2008).

If MPST is compared to classic Sanger sequencing (SS) it turns out that the new methods (costs: $1-60 per megabase) are much more economical than SS (costs: $ 0.50 per kilobase) (Shendure et al. 2008). Admittedly the quality of raw data is relatively low if compared to SS which has a per-base accuracy of 99,999%. Another disadvantage of MPST is the maximal read length of 400bp while a read length of up to ~1000 bp can be reached with SS technology.

Many bioinformatic tools for alignment, assembly, base calling and variant detection are available for all products which use MPST. The choice of the proper tool is an important measure to get adequate results. Mapping tools differ in mapping speed and accuracy. Different program features of available programs like CLC NGS Cell (http://www.clcbio.com), BLAT (Kent 2002), SSAHA2 (Ning et al. 2001), Bowtie (Langmead et al. 2009), SeqMap (Jiang 2008) and MAQ (Li et al. 2008) were compared by Palmieri and Schlötterer (Palmieri 2009).

I assume that a large number of variations in the used lines of my study can be found because flies of a whole population are used. I also expect divergence between reads and the reference genome because there is no available reference for *D. mauritiana*. This is the reason why I have to use the genome of *D. simulans* as reference. The most reliable mapping software detecting the absence or presence of indels and divergences between reads and reference are CLC and SSAHA2 (Palmieri 2009). Other programs don't provide gaps required in case of a missing base when mapping against the reference genome or they need high similarity between reads and the reference genome. Based on these considerations it proved necessary to decide for the bioinformatics tool CLC.

1.3. Project description

The *Drosophila simulans* complex is formed by three closely related species: *D. simulans, D. sechellia* and *D. mauritiana*. Previous studies showed that their genomes are homosequential and share variation (Kliman et al. 2000). While male progeny of interspecific crosses between the species mentioned above are infertile, the females are fertile and can backcross with males of any of the parental species (Haldane's rule). Due to their recent divergence, which is estimated to have occurred ~250.000 years ago (Hey et al. 1993; McDermott et al. 2008) it is supposed that gene flow might have occurred between these species. In fact, mitochondrial genome data seem to support this hypothesis (Ballard 2000). However the nuclear genome is much bigger and putatively introgressed regions might be difficult to find. With the emergence of MPST researchers have got a key tool to answer this question nowadays. We will be able to obtain a whole genome population data from *D. mauritiana* and use allele frequencies to estimate the proportion of *D. simulans* alleles introgressed into *D. mauritiana*.

The aim of this thesis is to validate data generated with the Genome Analyser II (GA II) by comparing SNPs. To determine the accuracy of population SNP frequencies which are supplied by Illumina data, I used SS technology to sequence two genomic regions (3L and X1) in individuals of the same lines which were also included in the Illumina dataset. My dataset, which I am going to call Sanger dataset, provides the "real" allele frequencies for those two regions in the *D. mauritiana* population pool. The Illumina dataset consisted of short reads (length ~50bp). I mapped the reads against a reference genome using the bioinformatics tool CLC. As there is no available genome for *D. mauritiana* which I could use as reference genome I used the genome of *D. simulans* as reference. This genome is known not to be of high quality. The difficulty was to map short reads against this reference genome avoiding lots of false negative SNPs caused by a big number of mismatched reads. I had to determine best mapping and SNP detection parameter combinations.

2. Materials and methods

2.1. Population samples

2.1.1. Samples of the Illumina pool

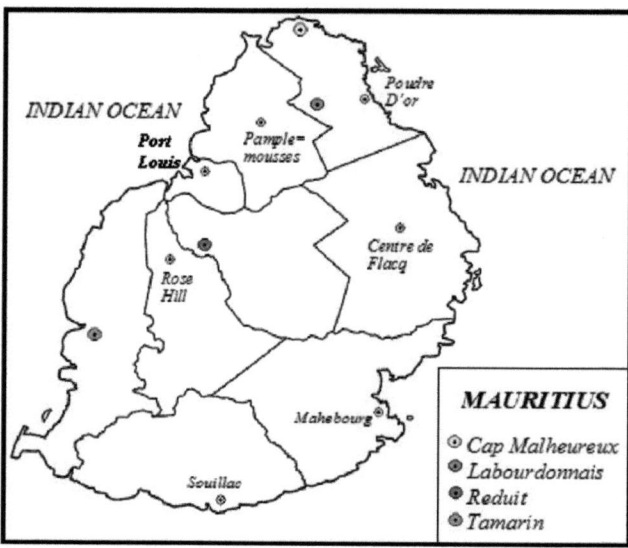

Figure 1: Map of Mauritius, including most populated cities (highlighted with black circles) and sample collection places of *D. mauritiana* (highlighted with colored circles).

As shown in figure 1 four populations of *D. mauritiana* were available for this study: In the case of the Reduit population our lab had access to inbred isofemale lines, kept in the lab since 2006. The other populations are virgin F1 females of *D. mauritiana* collected in Tamarin, Labourdonnais and Cap Malheurex. Because of the tendency of isofemale *Drosophila* lines to lose heterozygosity due to inbreeding, the contribution for the Illumina pool would be on average only one chromosome per Reduit line. As a pool strategy (all genomes were analyzed in one go) was chosen for working with the GA II, the 50 available lines of *D. mauritiana* from Reduit were crossed with each other in pairs to restore their heterozygosity. Our Illumina pool consists of virgin female progeny of the crossed Reduit lines (25 flies, one fly per cross) and 29 virgin F1 females of *D. mauritiana* collected in Tamarin, Labourdonnais and Cap Malheurex. After having been caught and stored in the laboratory the flies were frozen and kept at -80 °C until their DNA was extracted. Colleagues used the Qiagen kit for DNeasy Blood& Tissues for DNA extraction. All 54 flies were quickly pooled in a screw tube and crunched with a plastic pestle after immersion in liquid nitrogen. By using this pool strategy we get information about the absolute number of alleles and so also benefit from the cost-efficiency. The task of sample preparation wasn't carried out by me. When I started my project the reads of GA II were already available.

2.1.2. Samples used for Sanger sequencing

For Sanger sequencing I used DNA, extracted from individuals of the same lines of *D. mauritiana* which were used for Illumina sequencing. In the case of Reduit I used Fn males of each of the parental lines used in the Illumina pool. In the case of the lines from Tamarin, Labourdonnais and Capmalheureux I used F1 males which were kept in 96% ethanol.

2.2. DNA Preparation and extraction

Single male flies have been used for the DNA extraction. To distinguish them from the female ones I examined body size, sex combs in the first thoracic leg and genital shape with a stereoscope. The DNA high salt extraction protocol I used is described in the Appendix (Miller et al. 1988). I resuspended the DNA pellet in 30μl MilliQ water (Millipore) and eliminated RNA by using 1μl of RNAse (0.5μg/mol, Lactan) per sample. Agarose gel electrophoresis was done to check if the DNA extractions were successful. For SS I was looking for several 500-600bp regions of the genome of *D. mauritiana* that harbor SNP and indel polymorphism. The core region to be examined must be polymorphic but also flanked by conserved regions from where the primers can be designed. Therefore I chose introns of 350-500bp, flanked by 150bp exonic sequences. The sequences were obtained from a *D. simulans* dataset created by E. Meduri and primers were designed using Primer3, avoiding self priming and hairpin formation. To find out if the primers have unique genomic sequence and are indeed in conserved regions and if the core region examined contains interspecific site and indel differences I blasted them against the genomes of *D. melanogaster*, *D. simulans* and *D. sechellia* (http://flybase.org/blast/).

2.3. PCR, purification, quantification, sequencing reaction

Chromosome	Primer name	Primer sequence	Annealing temp.	Product size
X	*X_15266214*	FW: 5'TTATGCAATT CGTGGTCAGC-3'	55°C	592
	X_15265611	RV: 5'-TACATAGC CAAATGCCCTCC-3'		
3	*3L_15561972*	FW: 5'-GCGGAAGCT CAAGTTCTTTG-3'	55°C	606
	L_15562616	RV: 5'-GCTCCGTTA CCAGTACAGCC-3'		

Table 1: List of primers used for Sanger sequencing

In table 1 primers are listed that were chosen for Sanger sequencing. PCR was applied to amplify extracted DNA. Standard conditions are shown in table two.

Cycle step	Temp./°C	Time	Cycles
Initial denaturation	95	3 min	1
Denaturation	95	30 sec	35
Annealing	55	30 sec	35
Elongation	72	1 min	35
Final elongation	72	7 min	1

Table 2: Standard conditions for PCR

96Well Filter Plates (Millipore) were used for the purification of PCR products. I estimated the quantity of the products with agarose gel electrophoresis on a 0.8% agarose gel using the pBR322DNA/AluI Marker, 20 (MBI). For initiating the sequencing reaction I added 4-8μl DNA depending on the estimated concentration and MilliQ-water to a final volume of 10 μl. After the sequencing reaction was finished Sephadex G-50 Superfine System (GE Healthcare) and Multi-Screen 96-

well plates (Kat.No MAHVN4510, Lot.No. FOKN84305) were used for purification of the sequencing products.

Generally speaking, gel filtration is the best method for removing salt and dye terminators or any other small molecule contaminants. In comparison with most ethanol precipitation methods it is faster, more reproducible and delivers a better sequencing quality.

2.4. Capillary electrophoresis, sequence data acquisition and sequence editing

The capillary sequencer MegaBACE500 separated the sequencing reaction products. The electropherograms were analysed with CodonCode Aligner 3.0.1 (CodonCode Corporation, Dedham, USA). In most of the cases, forward and reverse strand of each sequence were used to build a consensus sequence.

2.5. Data analysis and sequence alignment

2.5.1. Sanger dataset

The Sanger dataset consists of sequences of 3L region (located on the 3rd chromosome) and X1 region (located on the X chromosome). As I wanted to compare this dataset to F1 flies of the Illumina dataset it turned out to be necessary to inspect the zygosity of the flies used for Sanger sequencing. The X1 region didn't show heterozygosity which means that only one allele for every gene is present, but in the 3L region some samples were heterozygous for indels. Heterozygosity in indels causes editing problems, so for the purpose of SNP frequency estimation and general polymorphism and divergence analysis, the region of heterozygous indels in 3L region (position 60- 365) was not considered.

The forward and the reverse strand where read to the point where they became heterozygous and gaps were put in the heterozygous regions. A heterozygous sequence is shown in figure 12. All sequences were aligned with ClustalW, and the alignments were edited with MacClade4.06. 14

For analyses of polymorphism and divergence I randomly discarded one allele for each heterozygous site in all samples of D. *mauritiana* that were inbred. They should count as having only one chromosome due to the loss of alleles during inbreeding.

For analyses of polymorphism DnaSP version 5.00.07 was used (Rozas 2009).

Nucleotide divergence measures the distribution of probability of average number of sites (Fixed differences in a sample of DNA compared to another sample are termed sites.) (Tajima 1983; Hey 1991; Wakeley et al. 1997).

Haplotype diversity (Hd) is a measure that shows the amount of differences of haplotypes in a given population (Nei 1987).

The Mutation Parameter (Θ) can be calculated per gene or per site. Its variance depends on the recombination between sites/genes. Generally speaking, in an autosomal gene of a diploid organism theta (Θ) = $4N\mu$.

N= effective population size

μ= the mutation rate per nucleotide site

Watterson's Theta is based on the number of segregating (polymorphic) sites (Watterson 1975):

$\Theta = S/a$

S= segregating sites

a = Σ (1 / i) from i= 1 to n-1

Nucleotide diversity (π) stands for an average number of nucleotide differences per site in two randomly chosen sequences of a sample population (Nei 1987; Nei et al.1990).

Linkage disequilibrium (LD) describes a non random association of alleles which can be located at the same chromosome or at another (Hill 1986).

Tajima Test (D test)

A mutation can evolve randomly (so there is no effect on fitness and survival) or under a non- random process like selection or intergression. This test, developed by Tajima (Tajima 1989), has been introduced for supporting the assumption that mutations are selectively neutral (Kimura 1983). The D test is based on the differences between the number of segregating sites and the average number of nucleotide differences.

With the program splitstree version 4.10 I tested recombination of *D. mauritiana* for 3L and X1 region.

2.5.1.1. Allele frequency table generation

The Sanger sequencing dataset has been used for allele frequency table generation where I randomly discarded one allele for each heterozygous site in all samples of *D. mauritiana* which were inbred.

2.5.2. Illumina dataset

To get the raw data of the genome of *D. mauritiana* three runs were performed by GA II. As reads are short (run 1 and run 2 ~43bp, run3 ~50bp) I preferred mapping against a reference genome instead of doing de novo assembly. The genome of *D. mauritiana* is not available, so I used the genome of *D. simulans* as reference.

When DNA is extracted it is not only the DNA of the target individual but it may also contain any other foreign organism like for example *Wolbachia*. This is a bacterium that is known to infect the cells of *D. mauritiana* (Rousset et al. 1995). If only the genome of the target individual is provided for the reference mapping, many of the reads corresponding to these foreign organisms like *Wolbachia* will be forced to map to the reference by the alignment program. To avoid that, I added three specific strains of *Wolbachia* to the reference genome. As the bacteria of the *D. mauritiana* have not been sequenced until now it seemed to be most promising to replace them by strains of closely related species.

Furthermore I added three different mitochondrial genomes to provide mitochondrial genomes that I expect to be present in the lines. In table 3 details are shown for three *Wolbachia* genomes and mitochondrial genomes added to the reference genome *D. simulans*.

Wolbachia genome	Genbank number (NCBI)
WB (*Culex quinquefasciatus*)	NC010981
WRi (*D.simulans*)	NC012416
Whel (*D.melanogaster*)	NC002978
Mitochondrial genome	
D. mauritiana strain (BG1)	AF200831
D. mauritiana strain (G52)	AF200830
D. melanogaster strain (Zimbabwe 53)	AF200829

Table 3: Details about *Wolbachia* and mitochondrial genomes added to the reference genome.

2.5.3. CLC

2.5.3.1. Mapping parameters

For analyses of raw data I worked with CLC Genomics Workbench 3.5.1. I set up different mapping parameter sets because I didn't know any divergence between the reads and the used reference genome of *D. simulans*. If there is much divergence I need a set that allows mismatch between reads and reference, but then I also have to expect a higher number of falsely mapped reads. The consultation of other lab members revealed, that it seemed favorable to set up one short and two long parameter sets for each region:

• SNPval1 10.6: long parameter set, proportion of reads that are allowed to mismatch the reference = 40%.

• SNPval3 10.8: long parameter set, proportion of reads that are allowed to mismatch the reference = 20%.

• SNPval4 : short parameter set, default values.

Using the 'contiguous report' option available in CLC I compared these three sets. In the contiguous report information about 'reference length', 'consensus length', 'mean coverage of reads', 'zero coverage regions', 'number of assembled reads', and 'non-specific matches' are listed. The 'consensus length' results from the mapped reads. CLC builds a consensus line automatically. The 'mean coverage of reads' describes how many reads map to the reference in average. 'Zero coverage regions' are regions, where no read maps. The 'number of assembled reads' counts all mapped reads. 'Non-specific matches' are mappings of reads that could map in other places too. In this case the program CLC chooses the place of mapping randomly.

2.5.3.2. SNP detection parameters

To find out true SNPs in mapped reads and to avoid that read errors falsify SNP results I have to set up SNP detection parameters in CLC. SNP detection was carried out by one long parameter set (SNPval3 10.8) and by the short parameter set. For both datasets, SNP detection was performed using two different sets of detection parameters:

- A minimum coverage of 10 reads, maximum coverage of 50 reads, a minimum allele frequency of 10% (or two reads) for regions X1 and 3L.
- A minimum coverage of 10 reads, maximum coverage of 50 reads, a minimum allele frequency of 20% (or two reads) for regions X1 and 3L.

3. Results

3.1. Population genetics analysis of *D. mauritiana*

3.1.1. Recombination

As recombination is a major drive of genetic variation that influences sequence variability and selection I want to know the recombination rate of the two regions 3L and X1. To look for evidence of recombination I made the Phitest (Bruen et al. 2006). It found statistically significant evidence for recombination (p=0.018) in the 3L region, but no evidence of recombination in the X1 fragment (p=0.224). I also calculated linkage disequilibrium. In figures 2 and 3 linkage disequilibrium (LD) is shown for both regions: the fragment X1 shows more outliers for the correlation between LD and distance.

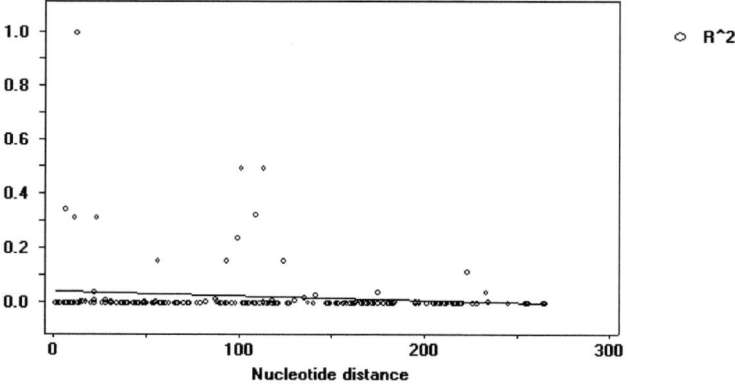

Figure 2: Graph of linkage disequilibrium calculated with R2, region 3L

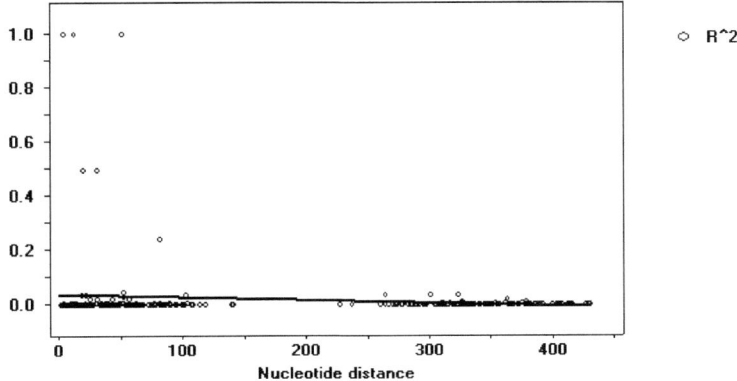

Figure 3: Graph of linkage disequilibrium calculated with R2, region X1

3.1.2. Nucleotide divergence and diversity

As shown in table 4 there is a higher value of D in the comparison between $D.$ *mauritiana* and $D.$ *melanogaster* as between $D.$ *mauritiana* and $D.$ *simulans* in 3L and X1 region.

	3L region	X1 region
D (mau-mel)	0.09909	0.04467
D (mau-sim)	0.02253	0.03605

Table 4: Shows the divergence between D *.mauritiana* and $D.$ *melanogaster* (D mau-mel) and $D.$ *mauritiana* and $D.$ *simulans* (D mau-sim).

In table 5 nucleotide diversity calculations are shown for 3L and X1 region. In 3L region, the value of *Hd* presents a similar amount of haplotypes as in X1 region. S, π, Θ (π) and Θ (S) values are higher in 3L region.

	Hd	π	Θ(S)	Θ(π)	S
3L region	0.7400	0.00534	0.02010	0.00538	25
X1 region	0.7080	0.00271	0.00892	0.00272	20

Table 5: Nucleotide diversity calculations of the two sequenced regions 3L and X1.

3.1.3. Tajima's Test (D test)

When the whole 3L region is analyzed, Tajimas'D is negative and shows a significant deviation from neutrality, as shown in table 6. For the synonymous sites and the silent sites (non coding parts of 3L region) and X1 region the analysis of Tajima's D is no longer significant.

	Tajima D	significance
3L region	-212.104	P < 0.05
synonymous sites	-0.52281	P > 0.10
silent sites	-130.247	P > 0.10
X1 region	-157.701	P > 0.10

Table 6: Tajima's D and statistical significance for 3L and X1 region.

3.2. Nucleotide divergence and diversity in comparison

I compare the results of nucleotide divergence and diversity of 3L and X1 region with other datasets that included the same species as in my study. To compare my results of Θ(S) with the results of Kliman et al. (2000), I put their results into a range with lower and upper limit, as shown in table 8. The mutation parameter of X1 and 3L is similar to their upper limit of high-recombination loci.

McDermott and Kliman (2008) subdivided their dataset in high and low recombination gene classes (HR, LR) before they calculated divergence (D). As seen in table 7 divergence between *D. mauritiana* and *D. melanogaster* in 3L region is roughly twofold compared to fragments they used. In X1 region it is almost equal to the value of D from high recombination (HR) and low recombination (LR) data. Low and high recombination genes show nearly the same divergence between *D. mauritiana* and *D. melanogaster*.

	3L	X1	HR, Klimann 2008	LR, Klimann 2008
D (mau-mel)	0.09909	0.04467	0.04050	0.04962
D (mau-sim)	0.02253	0.03605	0.01353	0.00046

Table 7: mau= *D.mauritiana*, mel=*D.melanogaster*, sim=*D.simulans*. Divergence (D), calculated for region L3, X1 (this study), HR and LR (McDermott et al. 2008).

	lower limit	upper limit
$\Theta(s)$ in low-recombination loci	0.0004	0.0023
$\Theta(s)$ in high-recombination loci	0.0031	0.0104
$\Theta(s)$ in X1		0.0089
$\Theta(s)$ in 3L		0.0201

Table 8: Range of $\Theta(s)$ in low and high recombination loci (Kliman et al. 2000), $\Theta(S)$ of regions X1, 3L.

3.3. Comparison between mapping runs using CLC report

To get comparable information about the 'consensus length', 'mean coverage of reads', 'zero coverage regions', 'all assembled reads' and 'non specific matches' the data of the report I got from CLC after every mapping run were put into a table (table 9). As shown in table 9 in 3L region the 'consensus length' varies from 20244629 bases (short parameter set), to 20867590 bases (long parameter set, 20%

mismatch allowed) to 20961014 bases (long parameter set, 40% mismatch allowed). The 'mean coverage of reads' counts ~47 reads using the long parameter set where 40% mismatch is allowed, ~43 reads using the more stringent long parameter set where 20% mismatch is allowed and ~31 reads using the short parameter set. 'Non-specific matches' vary from 279317 bases in the short parameter set to 444569 bases in the long parameter set where 40% mismatch is allowed. There are more 'non-perfect matches' in this long parameter set than in the other parameter sets. X1 region has a 'reference length' of 17042790 bases. The 'consensus length' is 14228629 bases using the long parameter set with an allowed mismatch of 40%, 14161088 bases using the long parameter set with 20% mismatch allowed, and 13813595 bases using the short parameter set. The 'mean coverage of reads' is similar to the results of 3L region. Most 'zero coverage regions' are counted using the short parameter set, which is also true for the 3L region. 'Non-specific matches' vary from 619019 bases (long parameter set, 40% mismatch allowed), to 526799 bases (long parameter set, 20% mismatch allowed) and 385831 bases (short parameter set). In the long parameter set where 40% mismatch is allowed most of non-perfect matches are counted.

	LPS, 40%	LPS, 20%	SPS	LPS, 40%	LPS, 20%	SPS
Reference length	22553184	22553184	22553184	17042790	17042790	17042790
consensus length	20961014	20867590	20244629	14228629	14161088	13813595
Mean coverage	46.97	43.06	30.92	47.17	42.86	30.78
Zero coverage	13542	15399	30045	11481	12931	23202
assembled reads	23289874	20601795	14187812	15944979	13929863	9639237
non-specific-matches	444569	378768	279317	619019	526799	385831
non-perfect-matches	19826156	17138234	10724356	13566145	11551278	7260392

Table 9: Contiguous report of 3L (*cursive*) and X1 region. LPS= long parameter set, SPS= short parameter set, percentage describes the allowed mismatch set up in parameter sets.

3.4. SNP Detection

As seen in table 10 in the long parameter set 14 SNPs (3L region)/ 20 SNPs (X1 region) were detected, in the short parameter set 5 SNPs (3L region)/ 11 SNPs (X1 region) were detected. Raising the minimal allele frequency from 10% up to 20% reduce the number of SNP in the short parameter set from 5 to 4 (3L region)/ from 11 to 10 (X1 region). To determine the number of true SNPs detected by parameter sets I compared these SNPs to SNPs which are detected by Sanger sequencing (table 11). In X1 region I detected 28 SNPs with the method of Sanger sequencing. The range of SNP frequency varies from 1.3% to 31.2% (only three SNPs have a frequency over 20%). The long parameter set marked 4 true and 16 false positive SNPs, so there are 80% false positive ones. 24 SNPs stayed undetected using the long parameter set. The short parameter set detected 2 true SNP, the number of false positive SNPs is much lower than in the long parameter set. About 80% are false positive SNPs. In the 3L region 24 SNPs were detected by Sanger sequencing. The range of SNP frequency varies from 1.3% to 24.4%. (only two SNPs appear in more than 20% of the lines). Using the long parameter set 14 SNPs were detected, 10 were false positive ones. So 71% are false positive SNPs. By using the short parameter set about 40% of SNPs were false positive.

	Set with a m.a.f. of 10% (or 2 reads)	Set with a m.a.f. of 20% (or 2 reads)
3L region		
LPS	14	14
SPS	5	4
X1 region		
LPS	20	20
SPS	11	10

Table 10: Comparison between total numbers of SNPs using different parameters. M.a.f.= minimal allele frequency.

	Total number of SNPs	true SNPs	% of true SNPs
3L region			
LPS	14	4	29%
SPS	4-5	2-3	50-60%
SS	24	24	100%
X1 region			
LPS	20	4	20%
SPS	10-11	2-3	20-27%
SS	28	28	100%

Table 11: Total number of SNPs and number of true SNPs detected with LPS (long parameter set), SPS (short parameter set) and SS (Sanger sequencing method).

4. Discussion

4.1. Population genetics analysis of *D. mauritiana*

4.1.1. Recombination

I expect high recombination rate in 3L and X1 region which are located towards the telomeres, because crossing over towards the telomeres takes place more often than towards the centromere (True et al. 1996). True and colleagues found also out that *D. mauritiana* has a higher recombination rate along the entire chromosomes than closely related species. I also calculated linkage disequilibrium (figures 2 and 3). LD shows little decrease with distance, as expected when high recombination occurs. The fragment X1 shows more outliers for the correlation between LD and distance which might explain the lack of statistical significance found with the Phitest.

4.1.2. Nucleotide divergence and diversity

I expect a higher value of divergence in the comparison between *D. mauritiana* and *D. melanogaster* than between *D. mauritiana* and *D. simulans* because *D. mauritiana* and *D. simulans* are much closer related to each other. As shown in table four results confirm with my expectations.

As shown in table 5 in 3L region, the value of *Hd* presents a similar amount of haplotypes as in X1 region. Furthermore it can be inferred that in 3L region more differences in diversity exist than in region X1, because π is higher in 3L region. Here, Θ (π) it is also higher than in X1 region. The larger number of polymorphic sites in 3L region also points out the higher nucleotide diversity.

4.1.3. Tajiama's Test (D test)

The 3L region includes a few codons from the exonic regions flanking the intron. When the whole 3L region is analyzed, Tajimas'D is negative and shows a significant deviation from neutrality, as shown in table 6. This indicates the action of purifying selection, as expected for coding regions. If I exclude the coding part of this fragment from the analysis Tajima's D is no longer significant. This means that the intronic part of the fragment is evolving neutrally. The X1 fragment contains only intronic sequence. Tajima's D shows no deviation from neutrality.

4.2. Nucleotide divergence and diversity in comparison with other dataset

As shown in table 8 the mutation parameter of X1 and 3L is similar to their upper limit of high-recombination loci what affirms that my regions are in high recombination. As expected, in both my regions divergence between *D. mauritiana* and *D. simulans* is similar to HR estimations from McDermott and Kliman.

4.3. Comparison between mapping runs using CLC report

In the course of the comparison I found out that the 'consensus length' gets longer and more accurate if parameter sets are used which allow many mismatches as in my study. The 'mean coverage of mapped reads' and the total 'number of assembled reads' also rise with the allowed number of mismatches. 'Zero coverage regions' which describe regions without any matched reads appear in a higher number in the short parameter set, which is the most stringent parameter set I used.

The less stringency in mapping parameters is set up the more 'non-specific' and 'non-perfect matches' are got.

4.4. SNP detection and examples

Comparing the two different SNP detection parameter sets, it is obvious that results are similar to each other (table 10). This is caused by the very low number of SNPs in my regions which have an allele frequency over 10%. The frequency of most of the SNPs ranges from 1.3% to 3.8%. The difference between SNP detection parameter sets is only one detected SNP, so my study shows no significant difference in using different SNP detection parameter sets. After comparing CLC reports of mapping runs (chapter 3.3.) I expect that the difference in SNP detection between mapping parameters is more important than using different minimal allele frequencies. Using the long parameter set two times more SNPs were recovered than with the short parameter set, but only ~1/5 out of them were true SNPs. This is caused by the high number of mismatches of reads in the long parameter set which produced false positive polymorphisms. The maximum of true SNPs recovered by CLC was 16% using the long parameter set. Regardless of the choice of parameter set the reason why lots of SNPs stayed undetected, is that the two regions consisted of a lot of SNPs which rarely appear. Because of the SNP detection parameter sets with a minimum allele frequency all these low frequency SNPs stayed undetected. All SNPs and the information about their positions, their allele variations, their frequencies and read coverage are shown in Appendix: Green accentuated single nucleotide polymorphisms are 'true' ones, also detected by classic Sanger sequencing. Red accentuated ones are false positive SNPs caused by mismatches. Shadowed SNPs are located in heterozygous regions. I exclude them from counting, because I cannot compare them with the Sanger sequencing

29

dataset where I exclude heterozygous regions (read more about it in chapter 2.6.1.). 24 SNPs stayed undetected using the long parameter set, which is mainly caused by the low frequency of polymorphism in the Sanger sequencing dataset. To point out the difficulty of true SNP detection I am going to show some special examples taken from the two regions 3L and X1 and discuss them.

X1 region:

X1 region consists of a lot of low frequency SNPs. As can be seen in table 12 SNPs which do not appear very often are not detected because of SNP detection parameters sets allowing SNPs with a frequency over 10 or 20%. Although differences in low frequency don't meet the minimum requirements to be detected they can be seen in the alignment of reads in many cases. As can be seen in figure 6, there is a 2bp insertion just in one of my lines (1.3% frequency) at osition 220. The long and the short parameter set harbor this insertion. Figure 5 shows the alignment window of the short parameter set.

Position	C	T	
position 190	97,4%	2,6%	not detected
position 198	1,3%	98,7%	not detected
position 433	98,7%	1,3%	not detected
position 426	1,3%	98,7%	not detected

Table 12: SNPs which are not detected.

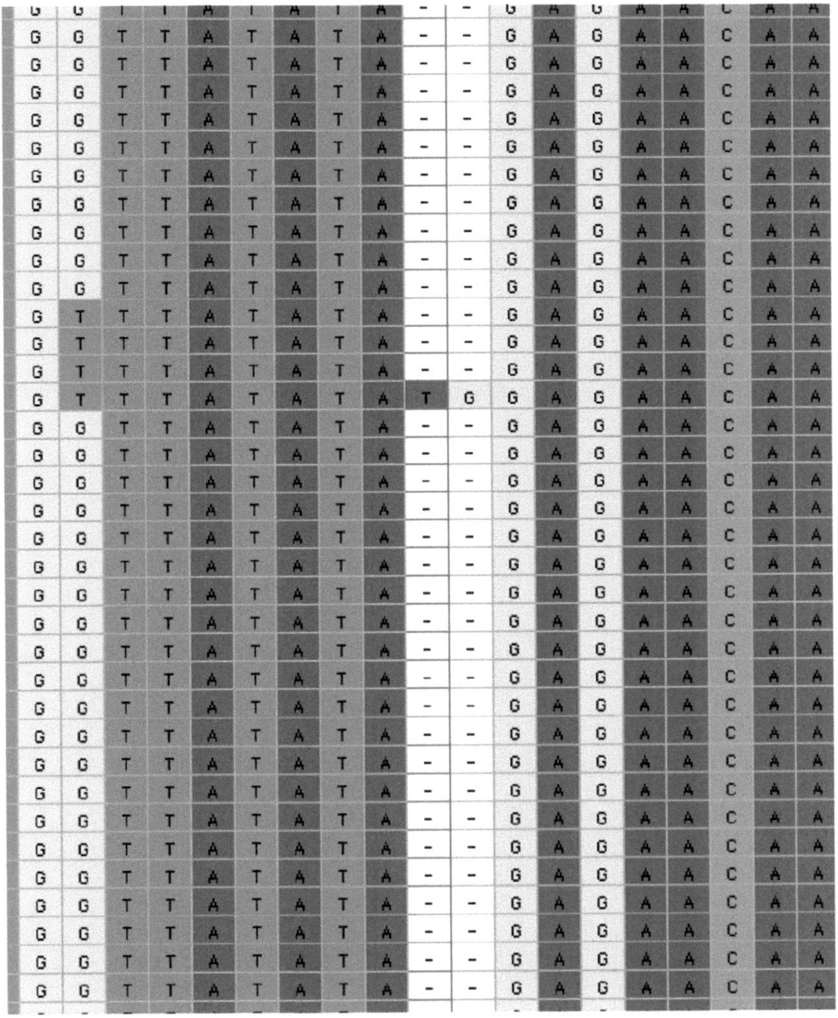

Figure 4: 2bp (T,G) insertion in one of the samples.

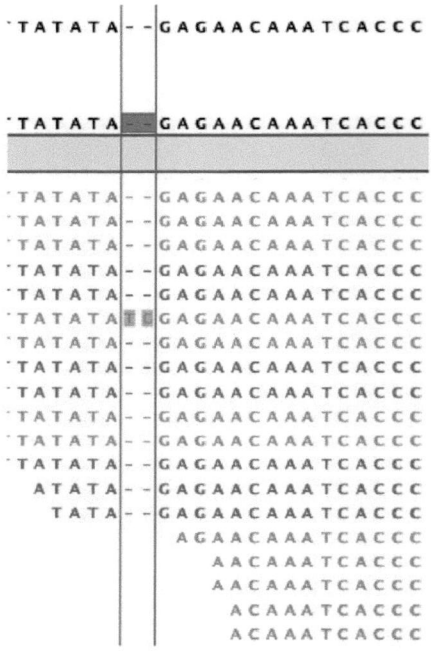

Figure 5: Typical alignment window of short parameter set. To determine if CLC works adequately with higher frequency variants I searched for an SNP with higher frequency and compared them. As shown in table 13, the SNP frequency of Sanger sequencing and the parameter sets used in CLC have similar results.

	A	G	% of A	% of G
Sanger sequencing	72	6	7,70%	92,30%
Short parameter set	28	2	6,70%	93,30%
Long parameter set	42	4	8,70%	91,30%

Table 13: Comparison of a 7.7% frequency SNP (position 246).

32

I searched a repetitive region (figure 5) to control the mapping of reads and SNP detection in such regions. In the long parameter set where mismatch is allowed an average number of reads map but often show impreciseness. A region of ~50bp is missing there - also in the consensus. In the short parameter set this region has low coverage of reads. Reads show the right polymorphism and SNPs are detected, even if they are in a repetitive region.

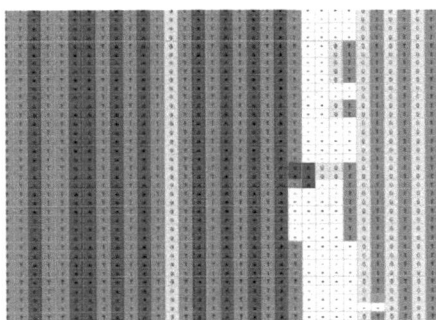

Figure 5: repetitive region, shown in McClade.

3L region

In the 3L region I deleted position 60- 365 in my Sanger sequencing dataset because of lines which are heterozygous for indel (read more in chapter 2.6.1.). As depicted in figure 6 and 7, the alignment cannot be managed correctly.

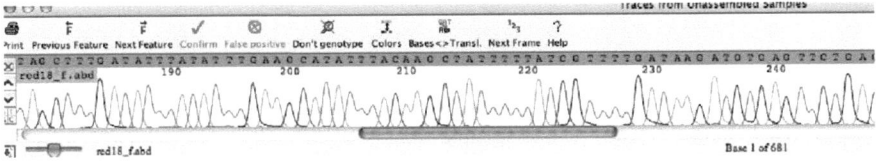

Figure 6: A homozygote sequence, shown in CodonCodeAlligner.

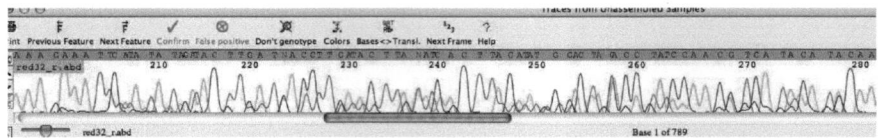

Figure 7: A herozygote sequence, shown in CodonCodeAlligner.

I manually looked at the reads in the heterozygous region to determine if there were troubles with mapping. In the long parameter set I found no decrease of mapped reads in the heterozygous region but as figure 8 shows in the short parameter set there is an explicit drop of reads.

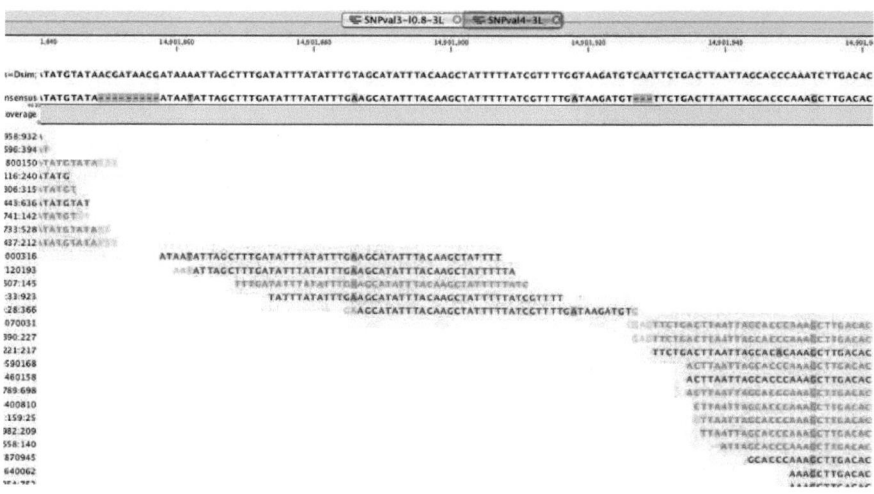

Figure 8: Alignment window of short parameter set.

4.5. Contamination challenge with *D. melanogaster*

After Illumina sequencing it was realized that the lines Red 6 and Red 31 kept in the lab were *D. melanogaster,* while they were originally classified as *D. mauritiana* (marked in Appendix with *). It is not known for sure when the contamination might have occurred and there is the possibility, though small, that these lines were included in the Illumina pool after the contamination. Therefore, I sequenced these *D. melanogaster* lines as well as the original Red 6 and Red 31 lines, which were recovered by asking colleagues to place their flies to our disposal (D. Hartl's lab). Lines Red 17, Red 20 and Red 57 (marked in Appendix with **) were also obtained from Hartl's lab because the flies were lost over time since the sequencing had been started with the GA II. These lines have the following labeling: HR 6, HR 17, HR 20, HR 31 and HR 56.

Besides the problem with the Reduit lines, we also observed a putative contamination in one line from Tamarin. In this study I first sequenced an F1 Tam 9 individual (marked in Appendix with ***) that had been put into Ethanol. This line should be *D. mauritiana* according to initial species determination carried out in the lab. However its sequence corresponds to a *D. melanogaster* variant. The Tam 9 isofemale line kept in the lab, as well as F2 progeny previously collected in Ethanol, only consists of *D. mauritiana* flies. Additionally I also found out that the F1 progeny of the vials brought from the field was a mixture of mostly *D. mauritiana* and few *D. melanogaster* individuals. A possible explanation for this is that the wild-caught female that gave rise to the Tam 9 line was probably brought together with a *D. melanogaster* female at the time of collection. I cannot be sure if the Tam9 individual sequenced by Illumina was a *D. mauritiana* or *D. melanogaster*, therefore I included both Tam 9 F1 *D. melanogaster* and a Tam 9 F2 individual in the dataset.

To determine the impact of putative contamination with *D. melanogaster* in the SNP frequencies estimated from the GA II data I assembled four different SS datasets:

- all *D. mauritiana* lines, except Tam 9 F2. Tam 9 (*D. melanogaster*) was included instead.
- all *D. mauritiana* lines, except HR 6 and HR 31. Red 6 and Red 31 were included instead (both *D. melanogaster*),
- all *D. mauritiana* lines , except Tam 9 F2, HR 6 and HR 31. Tam 9, Red 6 and Red 31 were included instead (*all D. melanogaster*),
- only *D. mauritiana* lines.

As table 14 shows, there is a high increase in number of SNPs when *D. melanogaster* is added to the dataset.

3L region	Amount of SNPs
SS (only *D.mau*)	24
SS (*D.mau*+ red6, red31)	40
SS (*D.mau*+ tam9)	40
SS (*D.mau*+ red6, red31, tam9)	40

X1 region	Amount of SNPs
SS (only *D.mau*)	28
SS (*D.mau*+ red6, red31)	45
SS (*D.mau*+ tam9)	46
SS (*D.mau*+ red6, red31, tam9)	47

Table 14: Comparison of SNPs in different contamination cases.

I looked for evidence of contamination with *D. melanogaster* by comparing *D. melanogaster* polymorphisms of SS with Illumina data: At position 408, 437 and

433 I found polymorphism between *D. mauritiana* and *D. melanogaster* in the Sanger sequencing dataset of 3L region. I manually examined the alignment in CLC for fixed differences between *D. melanogaster* and *D. mauritiana*. I found no evidence for contamination. I also searched for variants between *D. melanogaster* and *D. mauritiana* in data set of 3L region. As shown in table 15 only in one *D. melanogaster* sample (tam 9) a SNP is detected in the Sanger sequencing dataset. In the long and in the short parameter set this SNP is also present.

	T	G	mapped reads	perfectly matched
Sanger sequencing	98.70%	1.30%	--	--
long parameter set	87.20%	8.50%	47	11
short parameter set	89.50%	10.50%	38	9

Table 15: Comparision of SNP, position 38, 3L region

As shown in figure 9 in all three *D. melanogaster* samples an insertion is located, which starts at position 77. I don't find it in the parameter sets, because the reference genome of *D. simulans* which I used for the alignment in CLC doesn't harbor this insertion.

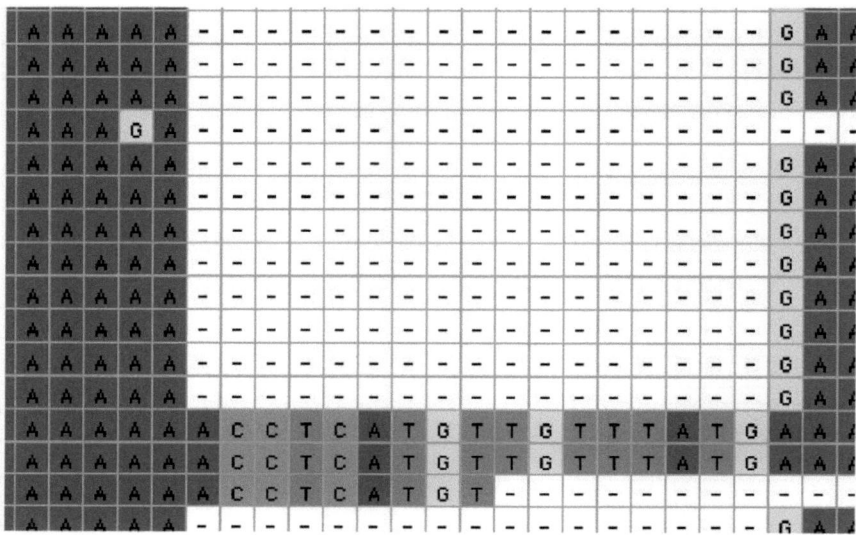

Figure 9: Insertion in *D. melanogaster* sequences

Due to these results it can be suggested that there is contamination with *D. melanogaster*.

4.6. Conclusions

One of the main conclusions of my study is that reads with a length of ~50bp are too short for detecting a large number of right positive SNPs. Mapping short reads is too inaccurate. Reads map in a more reliable way the longer they are.

A major problem in my study was the high number of false positive SNPs. Due to the matching of more reads a higher quality of consensus can be profited from, but more reads (which also means more non specific and non perfect matches) also increase the number of false positive SNPs. More stringency in mapping using the long parameter set would avoid many false positive SNPs. Results could also be improved by using quality scores to decrease the high number of false positive SNPs.

A lot of SNPs that are detected by Sanger sequencing don't appear in the SNP detection. This is caused by the high number of low frequency SNPs in the two regions sequenced by me. I checked both regions with the naked eye and I assert that reads often show the desired polymorphism. Due to the chosen parameters in SNP detection (a minimal of 10% and 20% SNPs) these low frequency SNPs don't appear in the SNP list. Decreasing the percentage may detect more low frequency SNPs.

An alternative to map against a reference genome is to do de novo assembly, where I expect an improvement of consensus compared to the used parameter sets. An advantage of this method is also achieved in the possibility of covering long insertions/ deletions.

In my study the best approach in SNP detecting with GA II would be to first make a less stringent run with a high acceptance of mismatches to get a good consensus. Finally a more stringent parameter for SNP detection is more efficient to reduce the high number of false positive reads.

4.7. Perspective

It is expected that the capacity of GA will rise again in the next years and which means that the costs will decrease even more. The fields of applications will become more the cheaper sequencing can take place, what furthermore results in an increasing importance.

Until now MPST has stretched across almost all fields of research with inconceivable pace (Teng et al. 2009). To get best results the proper method of massively parallel sequencing must be chosen. But that is not enough - parameter sets and quality values highly affect results. If parameters are balanced well, the GA offers opportunities in precision that are amazing.

The problem with mapping short reads against a reference genome in future won't exist, because technology proceeds with high speed and reads get longer. Nowadays it is possible to get reads up to 2x 75 bp and more, caused by the remarkable improvement of Genome AnalyserII only during the last year. Using 2x 75 bp for this study would yield much better results in SNP detection.

To enlarge this study, more different regions and quality values need to be tested with other parameter sets. A comparison with de novo assembly using the same lines would also be a very interesting task.

5. Appendix

5.1. SNP tables for Illumina data- X1 region

Reference position	Reference	Variants	Allele variations	Frequencies	Coverage of reads
11785228	C	2	C/G	95.1/4.9	41
11785230	T	3	T/A/G	88.6/4.5/4.5	44
11785233	G	2	G/C	95.5/4.5	44
11785235	T	2	T/G	93.5/4.3	46
11785367	C	2	C/T	91.3/8.7	46
11785372	A	2	A/G	89.5/5.3	38
11785396	C	2	C/G	90.5/9.5	21
11785403	C	2	C/A	78.3/13.0	23
11785414	T	3	T/A/G	84.6/7.7/7.7	26
11785415	C	2	C/A	92.3/7.7	26
11785476	T	2	G/T	81.8/18.2	11
11785518	G	2	G/T	90.0/10.0	40
11785519	G	2	G/T	87.8/12.2	41
11785540	A	2	A/C	92.5/7.5	40
11785543	T	2	T/G	94.9/5.1	39

Reference position	Reference	Variants	Allele variations	Frequencies	Coverage
11785615	A	2	A/G	68.2/31.8	22
11785617	A	2	A/G	50.0/50.0	22
11785622	A	2	A/G	52.4/47.6	21
11785665	C	2	C/A	86.7/13.3	30
11785678	A	2	A/G	94.3/5.7	35

Long parameter set, 20% mismatch allowed, 10% minimal allele frequency

Reference position	Reference	Variants	Allele variations	Frequencies	Coverage
11785230	T	2	T/G	86.7/6.7	30
11785238	G	2	G/T	94.4/5.6	36
11785244	G	2	G/T	94.9/5.1	39
11785261	C	2	C/A	95.8/4.2	48
11785281	A	2	A/C	94.0/4.0	50
11785367	C	2	C/T	93.3/6.7	30
11785519	G	2	G/T	88.6/11.4	35
11785599	G	2	G/A	90.0/10.0	10
11785615	A	2	A/G	60.0/40.0	10
11785617	A	2	A/G	70.0/30.0	10
11785622	A	2	A/G	80.0/20.0	10

Short parameter set,10% minimal allele frequency

Reference position	Variants	Allele variations	Frequencies	Coverage
11785228	2	C/G	95.1/4.9	41
11785230	3	T/A/G	88.6/4.5/4.5	44
11785233	2	G/C	95.5/4.5	44
11785235	2	T/G	93.5/4.3	46
11785367	2	C/T	91.3/8.7	46
11785372	2	A/G	89.5/5.3	38
11785396	2	C/G	90.5/9.5	21
11785403	2	C/A	78.3/13.0	23
11785414	3	T/A/G	84.6/7.7/7.7	26
11785415	2	C/A	92.3/7.7	26
11785476	2	G/T	81.8/18.2	11
11785518	2	G/T	90.0/10.0	40
11785519	2	G/T	87.8/12.2	41
11785540	2	A/C	92.5/7.5	40
11785543	2	T/G	94.9/5.1	39
11785615	2	A/G	68.2/31.8	22
11785617	2	A/G	50.0/50.0	22
11785622	2	A/G	52.4/47.6	21
11785665	2	C/A	86.7/13.3	30
11785678	2	A/G	94.3/5.7	35

Long parameter set, 20% mismatch allowed, 20% minimal allele frequency

Reference position	Variants	Allele variations	Frequencies	Coverage
11785230	2	T/G	86.7/6.7	30
11785238	2	G/T	94.4/5.6	36
11785244	2	G/T	94.9/5.1	39
11785261	2	C/A	95.8/4.2	48
11785281	2	A/C	94.0/4.0	50
11785367	2	C/T	93.3/6.7	30
11785519	2	G/T	88.6/11.4	35
11785615	2	A/G	60.0/40.0	10
11785617	2	A/G	70.0/30.0	10
11785622	2	A/G	80.0/20.0	10

Short parameter set, 20% minimal allele frequency

5.2. SNP tables for Illumina data- 3L region

Reference position	Variants	Allele variations	Frequencies	Coverage
14901748	2	A/C	96.0/4.0	50
14901750	2	T/G	87.2/8.5	47
14901764	2	C/G	93.3/4.4	45
14901766	2	G/T	95.1/4.9	41
14901770	2	C/A	94.7/5.3	38
14901773	2	T/G	86.8/7.9	38
14901780	2	T/G	94.1/5.9	34
14901787	2	A/C	93.8/6.2	32
14901789	2	C/T	85.7/11.4	35
14901790	2	C/A	94.1/5.9	34
14901823	2	A/C	90.0/10.0	20
14901877	2	T/G	90.0/10.0	10
14901885	2	G/T	90.0/10.0	10
14901917	2	G/T	90.0/10.0	20
14901935	2	A/T	88.9/7.4	27
14901955	2	T/C	90.3/6.5	31
14902020	2	T/C	87.5/12.5	16
14902075	2	T/A	78.9/15.8	19
14902097	2	A/G	77.3/18.2	22
14902149	2	G/T	68.4/31.6	19
14902167	2	C/A	92.5/7.5	40
14902174	2	G/C	93.0/4.7	43

Long parameter set, 20% mismatch allowed, 10% minimal allele frequency

Reference position	Variants	Allele variations	Frequencies	Coverage
14901750	2	T/G	89.5/10.5	38
14901789	2	C/T	86.2/10.3	29
14901790	2	C/A	92.9/7.1	28
14901823	2	A/C	89.5/10.5	19
14902097	2	A/G	80.0/13.3	15
14902118	2	T/G	90.0/10.0	10

Short parameter set, 10% minimal allele frequency

Reference position	Variants	Allele variations	Frequencies	Coverage
14901748	2	A/C	96.0/4.0	50
14901750	2	T/G	87.2/8.5	47
14901764	2	C/G	93.3/4.4	45
14901766	2	G/T	95.1/4.9	41
14901770	2	C/A	94.7/5.3	38
14901773	2	T/G	86.8/7.9	38
14901780	2	T/G	94.1/5.9	34
14901787	2	A/C	93.8/6.2	32
14901789	2	C/T	85.7/11.4	35
14901790	2	C/A	94.1/5.9	34
14901823	2	A/C	90.0/10.0	20
14901917	2	G/T	90.0/10.0	20
14901935	2	A/T	88.9/7.4	27
14901955	2	T/C	90.3/6.5	31
14902020	2	T/C	87.5/12.5	16
14902075	2	T/A	78.9/15.8	19
14902097	2	A/G	77.3/18.2	22
14902149	2	G/T	68.4/31.6	19
14902167	2	C/A	92.5/7.5	40
14902174	2	G/C	93.0/4.7	43

Long parameter set, 20% mismatch allowed, 20% minimal allele frequency

Reference position	Variants	Allele variations	Frequencies	Coverage
14901750	2	T/G	89.5/10.5	38
14901789	2	C/T	86.2/10.3	29
14901790	2	C/A	92.9/7.1	28
14901823	2	A/C	89.5/10.5	19
14902097	2	A/G	80.0/13.3	15

Short parameter set, 20% minimal allele frequency

References

Ahmadian, A., M. Ehn, et al. (2006). "Pyrosequencing: history, biochemistry and future." Clin Chim Acta **363**(1-2): 83-94.

Ballard, J. W. (2000). "Comparative genomics of mitochondrial DNA in members of the Drosophila melanogaster subgroup." J Mol Evol **51**(1): 48-63.

Bruen, T. C., H. Philippe, et al. (2006). "A simple and robust statistical test for detecting the presence of recombination." Genetics **172**(4): 2665-81.

Campbell, P. J., P. J. Stephens, et al. (2008). "Identification of somatically acquired rearrangements in cancer using genome-wide massively parallel paired-end sequencing." Nat Genet **40**(6): 722-9.

Chen, Y. T., M. Hsu, et al. (2009). "Cancer/testis antigen CT45: analysis of mRNA and protein expression in human cancer." Int J Cancer **124**(12): 2893-8.

Esteller, M. (2006). "The necessity of a human epigenome project." Carcinogenesis **27**(6): 1121-5.

Guffanti, A., M. Iacono, et al. (2009). "A transcriptional sketch of a primary human breast cancer by 454 deep sequencing." BMC Genomics **10**: 163.

Gupta, P. K. (2008). "Single-molecule DNA sequencing technologies for future genomics research." Trends Biotechnol **26**(11): 602-11.

Guryev, V. and E. Cuppen (2009). "Next-generation sequencing approaches in genetic rodent model systems to study functional effects of human genetic variation." FEBS Lett **583**(11): 1668-73.

Harismendy, O., P. C. Ng, et al. (2009). "Evaluation of next generation sequencing platforms for population targeted sequencing studies." Genome Biol **10**(3): R32.

Hey, J. (1991). "The structure of genealogies and the distribution of fixed differences between DNA sequence samples from natural populations." Genetics **128**(4): 831-40.

Hill, G. R., A (1986). "Linkage disequilibrium in finite populations." Theoretical and applied Genetics **38**: 226-231.

http://www.454.com/products-solutions/system-features.asp. **3.Juni 2009**.

http://www.clcbio.com. **16.Juni 2009**.

http://www.helicosbio.com/.**3.Juni 2009.**

http://www.illumina.com/. **3.Juni 2009**.

Ingman, M. and U. Gyllensten (2009). "SNP frequency estimation using massively parallel sequencing of pooled DNA." Eur J Hum Genet **17**(3): 383-6.

Jiang, H. and W. H. Wong (2008). "SeqMap: mapping massive amount of oligonucleotides to the genome." Bioinformatics **24**(20): 2395-6.

Kent, W. J. (2002). "BLAT--the BLAST-like alignment tool." Genome Res **12**(4): 656-64.

Kimura, M. (1983). "The neutral theory of molecular evolution." Cambridge University Press, Cambridge, Massachusetts.

Kliman, R. M., P. Andolfatto, et al. (2000). "The population genetics of the origin and divergence of the Drosophila simulans complex species." Genetics **156**(4): 1913-31.

Kozal, M. J. (2009). "Drug-resistant human immunodefiency virus." Clin Microbiol Infect **15 Suppl** 1: 69-73.

Langmead, B., C. Trapnell, et al. (2009). "Ultrafast and memory-efficient alignment of short DNA sequences to the human genome." Genome Biol **10**(3): R25.

Li, H., J. Ruan, et al. (2008). "Mapping short DNA sequencing reads and calling variants using mapping quality scores." Genome Res **18**(11): 1851-8.

McDermott, S. R. and R. M. Kliman (2008). "Estimation of isolation times of the island species in the Drosophila simulans complex from multilocus DNA sequence data." PLoS ONE **3**(6): e2442.

Miller, S. A., D. D. Dykes, et al. (1988). "A simple salting out procedure for extracting DNA from human nucleated cells." Nucleic Acids Res **16**(3): 1215.

Morozova, O. and M. A. Marra (2008). "From cytogenetics to next-generation sequencing technologies: advances in the detection of genome rearrangements in tumors." Biochem Cell Biol **86**(2): 81-91.

Nei, M. (1987). "Molecular Evolutionary Genetics." Columbia University Press, New York.

Nei, M. and J. C. Miller (1990). "A simple method for estimating average number of nucleotide substitutions within and between populations from restriction data." Genetics **125**(4): 873-9.

Ning, Z., A. J. Cox, et al. (2001). "SSAHA: a fast search method for large DNA databases." Genome Res **11**(10): 1725-9.

Odeberg, J., K. Holmberg, et al. (2002). "Molecular haplotyping by pyrosequencing." Biotechniques **33**(5): 1104, 1106, 1108.

Palmieri, N., Schlötterer, C., (2009). "Mapping Accuracy of Short Reads from Massively Parallel Sequencing and the Implications for Quantitative Expression Profiling." accepted in PLOS one (Juni 2009).

Ramaswamy, S., K. N. Ross, et al. (2003). "A molecular signature of metastasis in primary solid tumors." Nat Genet **33**(1): 49-54.

Rousset, F. and M. Solignac (1995). "Evolution of single and double Wolbachia symbioses during speciation in the Drosophila simulans complex." Proc Natl Acad Sci U S A **92**(14): 6389-93.

Rozas, J. (2009). "DNA sequence polymorphism analysis using DnaSP." Methods Mol Biol **537**: 337-50.

Schuster, S. C. (2008). "Next-generation sequencing transforms today's biology." Nat Methods **5**(1): 16-8.

Shendure, J. and H. Ji (2008). "Next-generation DNA sequencing." Nat Biotechnol **26**(10): 1135-45.

Shendure, J., R. D. Mitra, et al. (2004). "Advanced sequencing technologies: methods and goals." Nat Rev Genet **5**(5): 335-44.

Tajima, F. (1983). "Evolutionary relationship of DNA sequences in finite populations." Genetics **105**(2): 437-60.

Tajima, F. (1989). "Statistical method for testing the neutral mutationhypothesis by DNA polymorphism." Genetics **123**: 585-595.

Teng, X. and H. Xiao (2009). "Perspectives of DNA microarray and next-generation DNA sequencing technologies." Sci China C Life Sci **52**(1): 7-16.

Turcatti, G., A. Romieu, et al. (2008). "A new class of cleavable fluorescent nucleotides: synthesis and optimization as reversible terminators for DNA sequencing by synthesis." Nucleic Acids Res **36**(4): e25.

Vogelstein, B. and K. W. Kinzler (2004). "Cancer genes and the pathways they control." Nat Med **10**(8): 789-99.

von Bubnoff, A. (2008). "Next-generation sequencing: the race is on." Cell **132**(5): 721-3.

Wakeley, J. and J. Hey (1997). "Estimating ancestral population parameters." Genetics **145**(3): 847-55.

Watterson, G. A. (1975). "On the number of segregating sites in geneticalmodels without recombination. ." Theor. Pop. Biol. **7**: 256-276.

Wheeler, D. A., M. Srinivasan, et al. (2008). "The complete genome of an individual by massively parallel DNA sequencing." Nature **452**(7189): 872-6.

Wilhelm, B. T., S. Marguerat, et al. (2008). "Dynamic repertoire of a eukaryotic transcriptome surveyed at single-nucleotide resolution." Nature **453**(7199): 1239-43.

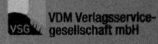

Printed by Books on Demand GmbH, Norderstedt / Germany